I0475244

Beach

COLORING BOOK FOR ADULTS

SKETCH DESIGN

COLORING BOOKS
FOR ADULTS

This Book
belongs to

- -

- -

- -

TEST YOUR COLOR

For Your
Design

For Your
Design

For Your
Design

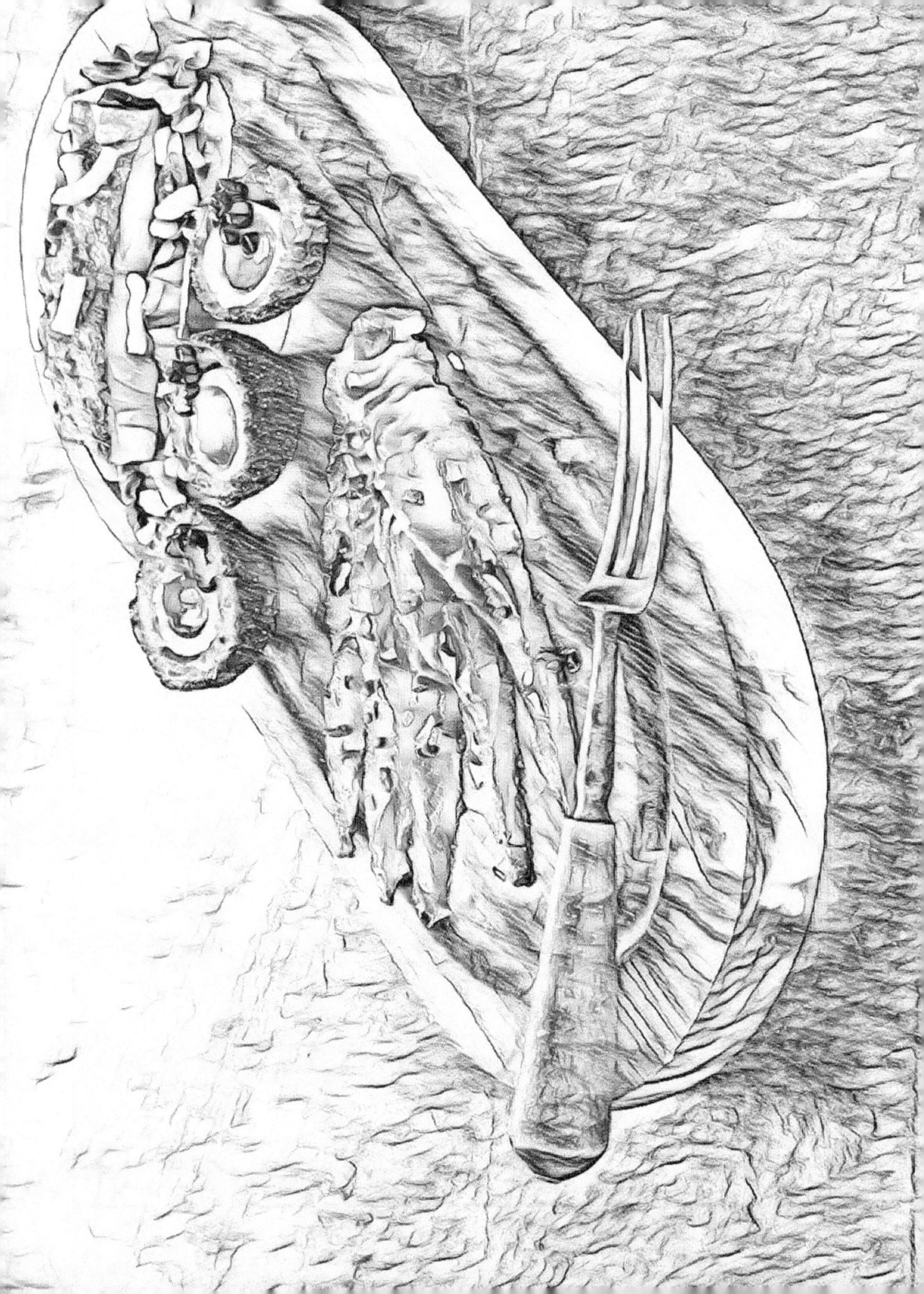

For Your
Design

For Your
Design

For Your
Design

For Your
Design

For Your
Design

For Your
Design

For Your
Design

For Your
Design

For Your
Design

For Your
Design

For Your
Design

For Your
Design

For Your
Design

For Your
Design

For Your
Design

For Your
Design

For Your
Design

For Your
Design

For Your
Design

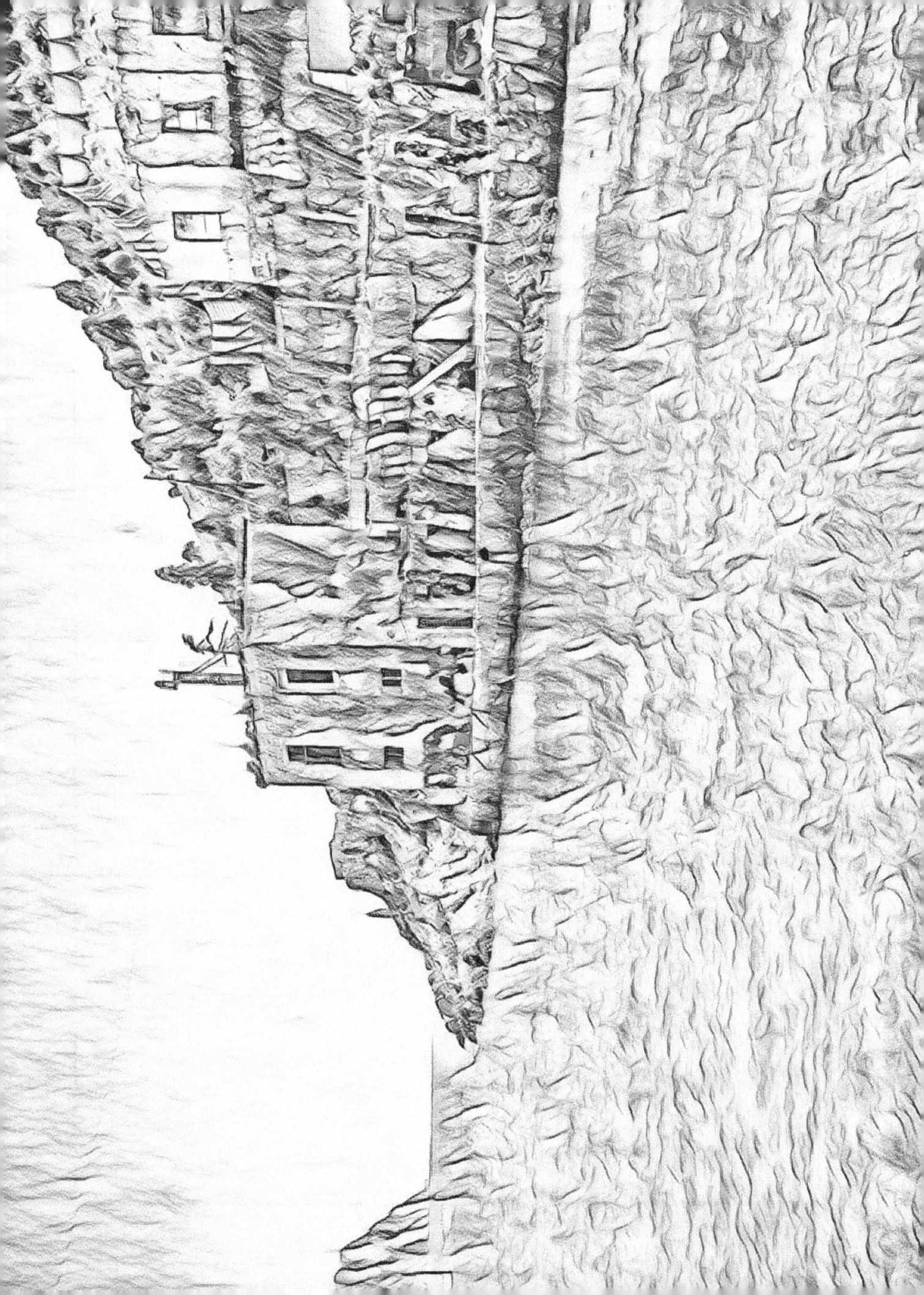

For Your
Design

For Your
Design

For Your
Design

For Your
Design

www.ingramcontent.com/pod-product-compliance
Lightning Source LLC
Chambersburg PA
CBHW081903170526
45167CB00007B/3131